The 6ft Edge

Your Guide to Thick Landscape Fabric

Terra Firma Royce

DEDICATION

This book is dedicated to those of you who find comfort in the arms of nature, whose hearts beat in time with the earth's pulse, and whose footsteps gently leave their marks on the ground.

I hope it will foster a closer bond, a love of stewardship, and lead us all to a day when environment and mankind coexist together.

With appreciation for your passion for the outdoors and with anticipation for future generations.

— Terra Firma Royce

CONTENTS

ACKNOWLEDGMENTS...1

CHAPTER 1..1

Overview of Sturdy Landscape Fabric................................1

    1.1 Heavy Duty Landscape Fabric: What Is It?.....................1

    1.2 Advantages of Using Thick Fabric..............................2

    1.3 When to Use Standard vs. Heavy Duty Fabric............4

    1.4 Selecting the Proper Weight and Material for the Fabric.........5

CHAPTER 2..8

Fabric Weight Understanding..8

    2.1 Oz (Fabric Weight Measurement): A Demystification.........8

    2.2 Typical Weights for Thick Landscape Fabric...............9

    2.3 Deciding on the Ideal Weight for Your Requirements.........11

    2.4 The Sturdiness and Lifespan of Various Weights.........12

CHAPTER 3...16

Uncovering the Strength of Sturdy Fabric.......................16

    3.1 Better Weed Control: Preventing Light and Growth.........16

    3.2 Improved Moisture Retention: Lowering Losses.........18

    3.3 Erosion Management: Safeguarding Gardens and Slopes.........19

    3.4 Suppressing Unwanted Plants and Ground Cover.........21

CHAPTER 4...24

Utilizing Heavy-Duty Fabric Applications........................24

    4.1 Driveways and Pathways: Establishing a Firm Foundation.........24

4.2 Mulch and Flower Beds: Keeping Your Eden Free of Weeds.......... 26

4.3 Vegetable Gardens: Increasing Produce and Reducing Weeds...... 28

4.4 Vineyards and Orchards: Controlling Weeds and Holding on to Moisture..29

CHAPTER 5... 32

Heavy Duty Fabric Installation Essentials................................. 32

5.1 Site Preparation: Clearing Out Previously Grown Plants and Debris.. 32

5.2 Laying the Material: Combining Overlapped Seams to Prevent Tears 33

5.3 Using Landscape Staples or Edging to Secure the Fabric................35

5.4 Plant Holes: Creating Straight Cuts for Simple Access....................36

CHAPTER N6..39

Combining Other Solutions with Heavy Duty Fabric...........................39

6.1 Mulch: Integrating Mulch and Fabric for Improved Weed Control and Aesthetics..39

6.2 Edging: Establishing Clearly Defined Boundaries and Avoiding Fabric Overflow... 41

6.3 Pre-Emergent Herbicide: An Intensive Method for Difficult Weeds... 43

6.4 Overlap of Landscape Fabric: Guaranteeing Total Weed Control... 45

CHAPTER 7... 49

Upkeep Advice for Durable Performance................................. 49

7.1 Examining for Tears and Punctures: Immediate Damage Management..49

7.2 Restocking Mulch: Sustaining a Layer That Prevents Weeds..........51

7.3 Eliminating Superfluous Weeds: Handling Breakouts Above the

Fabric..53

7.4 Using UV-Treated Fabric and Appropriate Installation to Increase

Fabric Lifespan..54

CHAPTER 8.. 58

Selecting the Ideal 6 feet wide landscape cloth...................................58

8.1 Benefits of a 6-foot-wide cloth: Reduced Seams and Simplified

Installation..58

8.2 Estimating Fabric Requirements for Your Project: Width and Depth.

60

8.3 Examining Custom Cut vs. Pre-Cut Options for 6-Feet of Fabric.... 62

8.4 Where to Get Premium 6 feet Wide, Thick Landscape Fabric........ 63

CHAPTER 9.. 67

Exceeding Weed Control: Surprising Advantages................................... 67

9.1 Controlling Temperature: Sustaining Lower Soil Temperatures..... 67

9.2 Play Areas: Establishing a Clean and Safe Environment for Kids.... 68

9.3 Pet Runs and Kennels: Reducing Mud and Upholding Sanitation.. 70

9.4 Building a Barrier to Slow the Spread of Flames: Firebreaks..........71

CHAPTER 10... 73

Using Heavy Duty Landscape Fabric to Go Green................................... 73

10.1 Eco-Friendly Choices for Recycled Content in Landscape Fabric.. 73

10.2 Options for Permeable Fabrics: Letting Water Pass Through....... 75

10.3 Biodegradable Fabric: Taking Future Disintegration Into Account...

76

10.4 Conscientious Fabric Use in Sustainable Landscaping: A

Responsible Approach...77

ABOUT THE AUTHOR... 81

ACKNOWLEDGMENTS

I want to express my sincere gratitude to everyone who helped make "The 6ft Edge " a reality.

I want to start by sincerely thanking my family for their constant support, tolerance, and understanding during this journey. Your support has been my compass during the creative process.

I express my gratitude to my editor and the publishing team for their belief in this project and their competent and dedicated guidance. This work has taken on its best form thanks to your insightful comments.

The issues covered in these pages are inspired by the persistent efforts of scientists, environmentalists, and environmental activists, to whom I am deeply obliged. Your dedication to protecting the environment is a ray of hope.

Last but not least, I would like to express my gratitude to all of the readers who have taken the time to read these words and explore our connection to the natural world. May a greater respect and care for the planet we call home be fostered by our common voyage.

With sincere appreciation,

Terra Firma Royce

Disclaimer

The opinions and viewpoints presented in this book, **"The 6ft Edge,"** are those of the author, Terra Firma Royce, alone, and may not represent the perspectives of any other person, group, or organization. The information here is solely for general informational purposes and is not meant to replace expert counsel. Regarding particular worries or situations, readers are urged to speak with pertinent authorities or professionals.

Amazon KDP Note:

It is my hope that this book, **"The 6ft Edge,"** will educate and excite readers about environmental issues and the wonders of nature. It doesn't contain anything that would elicit unfavorable comments or reviews from customers. The goal of the book is to encourage positive reader involvement and discussion by focusing on raising knowledge of and appreciation for nature.

I appreciate you taking the time to consider this.

Regards,

Terra Firma Royce

CHAPTER 1

1.1 Heavy Duty Landscape Fabric: What Is It?

For a variety of gardening and landscaping purposes, heavy duty landscape fabric is a durable, woven or non-woven cloth that offers long-lasting weed control and soil stabilization. Heavy duty landscape fabrics are designed to endure harder circumstances and offer longer performance than normal landscape fabrics, which are thinner and less resilient.

Important characteristics:

- **Material composition:** Heavy duty textiles, which are usually composed of polyester or polypropylene, are intended to withstand tearing and deterioration due to ultraviolet radiation.

- **Composite vs. Uncoated:** Interlaced strands give

1

woven clothes strength and allow water to move through while filtering out sunlight. Superior filtration and drainage are provided by the felt-like texture that is created when fibers are bonded together to create non-woven fabrics.

- **Weight and Thickness:** Compared to ordinary alternatives, heavy duty textiles are thicker and heavier, typically weighing three to five ounces per square yard or more.

1.2 Advantages of Using Thick Fabric

When it comes to gardening and landscaping jobs, high duty landscape fabric has many advantages over regular solutions.

Weed Control

- **Effective Barrier:** By physically blocking light from penetrating the soil surface, which is necessary for weed germination, the fabric inhibits the growth of weeds.
- **Decreases Upkeep:** It greatly lessens the

requirement for hand weeding and the application of chemical pesticides by preventing weed growth.

Soil Stabilization

- **Erosion Control**: By retaining soil particles in place, heavy duty fabric helps stabilize soil on slopes and in places that are prone to erosion.
- **Improves Soil Structure:** It helps maintain the structure of the soil, preventing compaction and encouraging the growth of healthy plant roots.

Moisture Retention

- **Water Conservation:** By lowering evaporation, the fabric aids in the retention of moisture in the soil, which is especially advantageous in arid conditions.
- **Efficient Irrigation:** This lessens the requirement for regular watering by ensuring that water reaches plant roots more effectively.

Durability and Longevity

- **Extended Lifespan:** Long-lasting materials with a

high degree of elasticity are made to resist UV radiation, high foot traffic, and inclement weather.

- **Economically Sound:** Their longevity and efficiency can eventually result in cost savings by lowering the need for maintenance and replacements, despite their higher original cost.

1.3 When to Use Standard vs. Heavy Duty Fabric

Selecting landscape fabric heavy duty or standard depends on the particular needs of your project.

Optimal Conditions for Sturdy Fabric

- **High-Traffic Areas:** Use on pathways, patios, and playgrounds areas with a lot of foot traffic.
- **Slope Stabilization:** When soil movement is an issue, slope stabilization is crucial for avoiding erosion.
- **Long-Term Projects:** Ideal for uses requiring long-term durability, like permanent plant beds, gravel pathways, and under driveways.

Situations in Which Standard Fabric May Be Sufficient

- **Temporary Installations:** Short-term projects or temporary installations where long-term durability is not essential can be satisfied with standard fabric.
- **Light Traffic Areas:** Ordinary cloth can effectively suppress weeds in gardens or other places with little foot traffic.
- **Cost Considerations:** Standard fabric might be a more cost-effective option for projects with a tight budget and little exposure to hostile environments.

1.4 Selecting the Proper Weight and Material for the Fabric

When choosing the right landscape fabric, it's important to take your project's weight, substance, and requirements into account.

Calculating Fabric Weight:

1. **Lightweight (1-2 ounces):** Ideal for light-duty uses where weed control is required but not a great deal

of soil stabilization, such vegetable gardens and flower beds.

2. **Medium Weight (3–4 ounces):** Excellent for most landscaping jobs, providing a good mix of robustness and manageability.

3. **Heavyweight (5+ ounces):** Ideal for tough situations requiring greatest strength and durability, such as driveways, slopes, and high-traffic areas.

Material Considerations:

1. **Polypropylene**: This material is a popular choice for long-term installations due to its strength and resistance to UV damage.

2. **Polyester:** Provides exceptional water permeability and durability; ideal for regions experiencing high relative humidity or frequent precipitation.

3. **Woven vs. Non-Woven:** Non-woven materials offer higher filtration and drainage capabilities, but woven materials give greater strength and longevity.

Other Considerations:

1. **UV Resistance:** Make sure the fabric is protected from deterioration by sunshine by adding UV inhibitors.
2. **Water Permeability:** To prevent waterlogging and promote plant health, select a fabric that permits sufficient water flow.
3. **Ease of Installation:** To help with accurate cutting and positioning, take into consideration fabrics with indicated lines or patterns.

You may choose a heavy duty landscape fabric that will last for a long time and operate at its best by being aware of the particular requirements of your project and its special qualities.

CHAPTER 2

FABRIC WEIGHT UNDERSTANDING

2.1 Oz (Fabric Weight Measurement): A Demystification

When evaluating whether landscape fabric is appropriate for a certain use, fabric weight is an important consideration. The material's density and thickness are commonly expressed in ounces per square yard, or oz/sq. yd.

Important Ideas:

1. **Ounces per Square Yard (oz/sq. yd)**: This measurement indicates the fabric's weight distributed over a square yard. Denser, thicker fabric is indicated by a higher weight.

2. **Density and Thickness:** Denser and thicker materials offer better strength, resilience to abrasion,

and life expectancy.

3. **Impact on Performance:** The weight of the fabric has an impact on its longevity, ability to stabilize the soil, and efficacy in controlling weeds.

In order to choose the best landscape fabric for your project and ensure lifespan and optimal performance, it is crucial to understand fabric weight.

2.2 Typical Weights for Thick Landscape Fabric

There are several weights available for heavy duty landscape fabrics, and each is appropriate for a particular use. The most popular weights consist of:

Small (1-2 oz/sq. yd)

- **Applications:** Suitable for gardening jobs that require little stabilization of the soil, such flower beds and vegetable gardens.
- **Features:** Lightweight and manageable, providing minimal weed control without appreciably changing soil moisture content.

3–4 oz/sq. yd. Medium Weight

- **Applications:** Suitable for garden walks, perennial beds, and areas with moderate traffic.
- **Characteristics:** Offers a blend of strength and flexibility, effectively controlling weeds and providing modest soil stabilization.

Heavyweight (5+ ounces per square yard)

- **Applications:** Suitable for hard projects requiring maximum strength and endurance, such as driveways, slopes, and high-traffic areas.
- **Characteristics:** Thick and robust, providing excellent soil stability, weed control, and resistance to tearing and UV damage.

Choosing the right weight guarantees that the landscape fabric works well in the intended application and offers long-term advantages.

2.3 Deciding on the Ideal Weight for Your Requirements

Selecting the appropriate fabric weight requires evaluating the particular needs of your landscaping project and comparing them to the characteristics of various fabric weights.

Considerations:

1. **Project Type:** Identify the type of project you're working on, such as a garden bed, pathway, or slope, and figure out how durable and stabilizing the soil it has to be.
2. **Soil Conditions:** Choose a fabric that can efficiently control the soil's characteristics, such as moisture content and erosion potential.
3. **Traffic Levels:** Take into account how much foot or car traffic the region may see. For increased durability, heavier textiles are needed in higher traffic regions.
4. **Climate Conditions:** Examine the UV exposure, precipitation, and temperature swings in the area to

select a fabric that will not deteriorate.

Real-World Examples:

- **Garden Beds and Borders:** For standard garden beds, a medium-weight (3–4 oz/sq. yd) cloth offers sufficient weed control and soil stabilization.
- **Driveways and Paths:** To ensure long-term durability, heavyweight (5+ oz/sq. yd) fabric is required to withstand the strain and wear from foot or vehicle activity.
- **Slope Stabilization:** In places where soil movement is likely, use a heavyweight fabric to keep slope integrity intact and stop soil erosion.

You may choose the ideal fabric weight for your project by carefully weighing these variables, which will ensure longevity and good performance.

2.4 The Sturdiness and Lifespan of Various Weights

The weight of landscape fabric has a direct impact on its resilience and lifetime. Generally speaking, heavier textiles function longer because they are more resilient to deterioration.

Fabrics that are lightweight (1-2 oz/sq. yd)

- **Durability:** These textiles have a basic level of durability, making them appropriate for light-duty or temporary uses. They are more vulnerable to deterioration and ripping.
- **Duration:** usually needs more regular replacement or maintenance after lasting one to two years in outdoor environments.

Fabrics with a medium weight (3–4 oz/sq yd)

- **Durability:** Offers a modest degree of resilience against tearing and UV rays, making it appropriate for routine landscaping duties.
- **Longevity:** Can endure three to five years with appropriate installation and upkeep, providing a decent trade-off between price and functionality.

Heavyweight Textiles (5+ oz/sq. yd)

Durability: Provides the highest level of resilience against rips, tears, and UV deterioration. Perfect for applications with a lot of traffic and demands.

Duration: can provide long-term performance and cost savings for up to 10 years, depending on the installation quality and environmental factors.

Increasing Durability:

- **Correct Installation:** Make sure the cloth is positioned appropriately, with overlap and anchoring in place to avoid moving and exposure.
- **Maintenance:** To ensure continued efficacy, periodically inspect the fabric for wear or damage, and replace or repair any damaged areas.
- **UV Protection:** To improve resistance to sunlight and extend the fabric's life, select fabrics with UV inhibitors.

Knowing how long and how durable various fabric weights are can help you make wise choices and guarantee that your landscape fabric investment will yield the best results over time.

CHAPTER 3

UNCOVERING THE STRENGTH OF STURDY FABRIC

3.1 Better Weed Control: Preventing Light and Growth

Strong landscape fabric is well known for its potent weed-suppression capabilities. Its capacity to block light and inhibit weed growth is mostly attributable to its shape.

Weed Suppression Mechanism:

1. **Blocking Sunlight:** The fabric keeps sunlight out of the soil, which is necessary for the germination and growth of weeds. Weed seeds require light to germinate.
2. **Physical Barrier:** Weed roots are prevented from entering the soil and emerging above ground by the dense structure of heavy duty fabric, which serves as a physical barrier.

Advantages:

- **Lower Maintenance**: Heavy duty fabric effectively inhibits weed development, reducing the need for hand weeding or chemical pesticide treatment.
- **Enhanced Health of Plants:** Desirable plants have less competition for nutrients, water, and sunlight when weeds are reduced, which encourages healthy growth.

Installation tips:

- **Overlap Edges**: Make sure the fabric's edges overlap by a minimum of 6 inches to avoid weeds growing through any gaps.
- **Secure with Stakes:** To keep landscape fabric firmly attached to the ground and stop weeds from growing through, use stakes or staples.

Maintaining a neat and healthy environment may be done effectively and sustainably by using heavy duty cloth for weed suppression.

3.2 Improved Moisture Retention: Lowering Losses

The capacity of heavy-duty landscape fabric to hold onto soil moisture and minimize the need for frequent watering is one of its many noteworthy benefits.

Mechanism of Moisture Retention:

1. **Reducing Evaporation:** By acting as a barrier, the fabric helps to keep moisture in the soil for extended periods of time by reducing the amount of water that evaporates from the soil surface.
2. **Even Distribution:** It aids in the uniform distribution of water throughout the soil, guaranteeing that the moisture content of plant roots is constant.

Advantages:

- **Water Conservation:** Heavy weight fabric reduces evaporation, which helps preserve water and makes it an environmentally responsible option, particularly in desert areas.

- **Healthy Plant Growth:** Because plants receive a consistent supply of water, which is necessary for their development, consistent moisture levels encourage healthier plant growth.

Installation Advice:

- **Correct Positioning:** To optimize the fabric's capacity to retain moisture, make sure it is placed flat and in close proximity to the soil.
- **Mulching:** Covering the cloth with a layer of mulch can improve its ability to retain water and further minimize evaporation.

Retaining soil moisture with heavy-duty landscape fabric is a wise move in sustainable gardening and landscaping.

3.3 Erosion Management: Safeguarding Gardens and Slopes

A strong landscape fabric is a useful technique for preventing soil erosion, especially in gardens that are prone to soil displacement and on slopes.

Mechanism of Erosion Control:

1. **Soil Stabilization:** By retaining soil particles in situ, the fabric stabilizes the soil and keeps irrigation or rain from washing them away.
2. **Reinforcement:** It lowers the risk of landslides and soil movement on slopes by adding to the soil structure's support.

Advantages:

- Gardening and Landscaping Project Investments Are Protected: Heavy strength fabric helps preserve the integrity of planted areas by minimizing soil erosion.
- Stabilized soil creates a more stable and healthy growing environment for plant roots, which in turn promotes healthier plant growth.

Installation Advice:

- **Anchoring:** To guarantee the cloth stays in place and offers the greatest amount of erosion prevention, fasten it with stakes or staples, especially on slopes.

- **Overlap and Seams:** To keep soil from slipping through gaps and compromising the effectiveness of the fabric, overlap the edges of the cloth and tuck in any seams.

For slopes and garden areas to remain stable and healthy, erosion control with heavy-duty landscaping fabric is crucial.

3.4 Suppressing Unwanted Plants and Ground Cover

Heavy strength landscaping cloth works well at suppressing undesired ground cover and invasive species in addition to weeds.

Mechanism of Suppression:

1. **Physical Barrier:** Unwanted plants are kept from spreading and establishing roots in undesirable places by the fabric's physical barrier action.
2. **Limiting Growth:** The fabric prevents invasive ground cover from growing by obstructing sunlight

and reducing the amount of area available for root growth.

Advantages:

- **Managing Invasive Species:** Sturdy fabric aids in keeping invasive species under control so they don't trample attractive plants and landscaped areas.
- **Maintaining Aesthetics:** The fabric contributes to the preservation of gardens' and landscapes' aesthetic appeal by suppressing undesired ground cover.

Installation Advice:

- **Targeted Application:** Apply the fabric, making sure it covers the entire affected area, in locations where invasive plants and ground cover are an issue.
- **Periodic Inspection:** Check the fabric on a regular basis to make sure it's still strong and intact, and fix any tears or gaps right away.

To keep your environment well-managed and visually appealing, use heavy-duty landscape cloth to suppress undesired plants and ground cover.

CHAPTER 4

UTILIZING HEAVY-DUTY FABRIC APPLICATIONS

4.1 Driveways and Pathways: Establishing a Firm Foundation

Building driveways and walkways requires the use of heavy-duty landscaping fabric, which strengthens the foundation and extends the life of the surfaces.

Establishing a Stable Base:

1. **Soil Separation:** The fabric prevents soil from mixing with sand or gravel by acting as a barrier between the soil and the aggregate base material. A stable and long-lasting surface is ensured by this separation, which preserves the integrity of the foundation layer.

2. **Load Distribution:** To lessen the chance of rutting and surface deformation, heavy duty fabric helps

spread the weight of cars and foot traffic uniformly throughout the base.

Advantages:

- **Longer Lifespan:** The fabric prolongs the life of roads and paths by limiting soil movement and preserving the stability of the base, which lowers the frequency of repairs.
- **Lower Maintenance:** Over time, less maintenance tasks are needed since the fabric's strong foundation reduces the development of surface cracks and potholes.

Installation Advice:

- **Correct Positioning:** Spread the cloth out evenly across the entire area underneath the driveway or pathway on a level, well-prepared surface.
- **Securing Edges:** To keep the fabric from slipping while the aggregate base is being installed, secure the edges with landscape staples or pins.

The use of heavy-duty landscaping fabric for driveways and paths guarantees a surface that is sturdy, long-lasting, and low-maintenance, improving outdoor areas' usability and aesthetic appeal.

4.2 Mulch and Flower Beds: Keeping Your Eden Free of Weeds

For weed-free flower beds and mulch areas, heavy duty landscape fabric is a great way to encourage healthy plant growth and lessen care requirements.

Weed Control:

1. **Light Blockage:** By keeping sunlight from penetrating the soil, the fabric stops weeds from germinating and growing. This keeps flower beds weed-free and aesthetically pleasing.
2. **Physical Barrier:** Weed roots are prevented from entering the soil and emerging above ground by the fabric's dense structure, which serves as a physical barrier.

Advantages:

Healthy Plants: By removing weed competition, desirable plants are better able to obtain sunlight, water, and nutrients, which leads to healthier growth.

Decreased Vegetation: Time and effort are saved because of the fabric's efficient weed suppression, which reduces the need for hand weeding.

Installation Advice:

- **Fabric Coverage:** Make sure the fabric completely encloses the flower bed, overhanging the edges by a minimum of 6 inches to keep weeds from growing in the spaces between.
- **Mulch Layer:** To boost the fabric's ability to suppress weeds and the flower bed's visual appeal, add a layer of mulch on top of it.

A stunning, low-maintenance garden area that promotes healthy plant growth can be created by using heavy duty landscape fabric in flower beds and mulch areas.

4.3 Vegetable Gardens: Increasing Produce and Reducing Weeds

Heavy-duty landscaping fabric is essential for vegetable gardens because it increases yields while lowering maintenance and weeding costs.

Boosting Vegetable Growth:

1. **Moisture Retention:** By lowering evaporation, the fabric helps keep soil moisture, ensuring that vegetable plants receive a steady supply of water, a necessity for their development and productivity.
2. **Weed Suppression:** The fabric physically blocks sunlight and acts as a barrier to effectively stop weed growth, hence lowering competition for nutrients and space.

Advantages:

Higher Yields: Vegetable plants can grow more vigorously and produce higher-quality food as weeds aren't fighting for the same resources.

Streamlined Upkeep: Gardeners may concentrate on plant care and harvesting because there is less need for weeding and watering, which makes garden upkeep easier.

Installation tips:

- **Planting Holes:** Make sure the holes in the fabric are just big enough to fit the vegetable plants without exposing too much soil.
- **Correct Anchoring:** To keep the cloth in place and avoid shifting during planting and watering, fasten it with garden staples.

In vegetable gardens, heavy-duty landscape fabric enhances growth conditions, resulting in healthier plants and abundant harvests that require little care.

4.4 Vineyards and Orchards: Controlling Weeds and Holding on to Moisture

Heavy-duty landscape fabric plays a crucial role in keeping soil moisture and controlling weeds in orchards and vineyards, hence enhancing the general well-being and yield of fruit-bearing plants.

Moisture Retention and Weed Suppression:

1. **Weed Control**: By keeping weeds out of the way of orchard and vineyard plants' access to nutrients, water, and sunlight, the fabric encourages healthier growth and larger yields.

2. **Moisture Conservation:** The fabric helps preserve soil moisture by lowering evaporation, which guarantees fruit trees and vines a consistent flow of water—a necessary component for fruit development.

Advantages:

- **Enhanced development:** By suppressing weeds and retaining moisture, an ideal environment is created for plants that bear fruit, which promotes enhanced development and productivity.

- **Reduction in Work:** Because of the fabric's efficient weed control, less time and money are wasted on labor-intensive manual weeding and chemical treatments.

Installation Advice:

- **Wide Coverage:** To get the most out of the fabric, place it in wide strips between rows of trees or vines, being sure to cover the entire root zone.
- **Securing Edges:** To stop wind or water from moving the fabric, firmly anchor it with landscape staples or pins.

The implementation of robust landscape fabric in orchards and vineyards fosters a more effective and fruitful growth atmosphere, hence augmenting the total prosperity of fruit production enterprises.

CHAPTER 5

5.1 Site Preparation: Clearing Out Previously Grown Plants and Debris

Installing high strength landscaping fabric successfully requires careful site preparation. By taking this step, you can make sure the fabric stays effective over time and lies flat.

Site preparation steps:

1. **Clear the Area:** Take out any existing grass, weeds, and other plants. To make sure there are no plants remaining in the area, use a garden hoe, shovel, or weed trimmer.

2. **Remove Debris:** Get rid of any pebbles, sticks, or other objects that can rip or puncture the cloth. To provide the cloth a level foundation, level out the

soil's surface.

3. **Equalize the Terrain:** Level the ground by removing high places and filling in low spots for optimal results. This guarantees that after installation, the fabric will lie flat and won't bunch up or move.

Importance of Careful Preparation:

- **Preventing Tears**: Leveling the ground and removing sharp objects helps keep the fabric from tearing and puncturing, preserving its effectiveness.
- **Enhanced Weed Control:** Complete vegetation removal maximizes the fabric's ability to suppress weeds by preventing weeds from growing back through it.

The basis for an effective and long-lasting landscape fabric installation is proper site preparation.

5.2 Laying the Material: Combining Overlapped Seams to Prevent Tears

For the fabric to effectively suppress weeds, retain moisture, and control erosion, it must be laid appropriately.

Procedure for Fabric Layout:

1. **Measure and Cut:** Determine the area that has to be covered, then cut the fabric to fit, leaving room for some overlap. To make precise cuts, use a utility knife or a pair of sharp scissors.
2. **Seam Overlap:** Make sure the fabric's edges are at least 6 inches overlapped when covering a broad area. This keeps weeds from growing through the spaces left by the fabric's rips.
3. **Avoiding Tears:** Take care when handling the fabric to prevent tearing. Smooth the fabric to ensure it rests flat against the ground and to get rid of any creases before placing it.

Importance of Correct Laying:

- **Seam Integrity:** Overlapping seams guarantee a continuous barrier against soil erosion and weeds, preserving the fabric's efficacy.

- **Durability:** The fabric's structural integrity is preserved and its lifespan is extended by careful handling and seamless installation, which averts tearing.

A crucial step in determining the overall effectiveness and longevity of the landscape fabric installation is properly placing the fabric.

5.3 Using Landscape Staples or Edging to Secure the Fabric

To stop the cloth from moving because of wind, water, or foot activity, it must be secured in place.

Techniques for Fastening the Fabric:

1. **Tree Staples:** To fasten the fabric to the ground, use fabric pins or landscape staples. Staples should be positioned at strategic locations in the fabric's center and every 12 to 24 inches along the edges and seams.

2. **Corner:** To further secure the cloth, think about

adding landscape edging in addition to staples. Edging gives the fabric a polished appearance and more support, keeping it from shifting.

Significance of Appropriate Fastening:

- **Avoiding Motion:** Fastening the cloth firmly keeps it from moving, which can leave spaces for weeds to proliferate and lower the fabric's overall efficacy.
- **Long-Term Stability:** Fabric that is securely fastened remains in situ over time, offering reliable erosion protection, moisture retention, and weed control.

Securing the fabric with landscaping staples and edging guarantees a sturdy and long-lasting installation that withstands weather conditions.

5.4 Plant Holes: Creating Straight Cuts for Simple Access

To grow and manage flora while preserving the fabric's benefits, plant holes must be cut in the cloth.

Cutting Plant Holes:

1. **Mark the Locations:** Mark the areas where plants will be positioned prior to cutting. Make sure the plants being put are spaced appropriately.

2. **Make Neat Incisions:** To create precise cuts in the fabric, use scissors or a sharp utility knife. Make a cut that is an "X" or a circle big enough to hold the plant root ball, but not so big that it is too open.

3. **Refold the Cloth:** To make a planting hole, gently fold back the fabric's cut parts. To reduce dirt exposure after planting, fold the fabric back around the base of the plant.

The Significance of Clean Cuts:

Minimizing Soil Exposure: Cutting carefully and folding back the cloth decreases soil exposure, which lowers the likelihood of weed growth surrounding the plant.

Plant Health: Tight cuts keep the fabric from tearing and guarantee that it stays intact, which helps the plants and the landscape as a whole over time.

Ensuring that the fabric stays functional while providing easy access for planting and maintaining vegetation is achieved by the accurate cutting of clean plant holes.

Heavy duty landscape fabric installation requires exact cutting for plant access, careful fabric placement, strong anchoring, and site preparation. In order for the fabric to properly suppress weeds, retain moisture, and reduce erosion and give gardens and landscapes long-term advantages, each stage is crucial.

CHAPTER N6

6.1 Mulch: Integrating Mulch and Fabric for Improved Weed Control and Aesthetics

Combining mulch and heavy-duty landscaping fabric is a great way to improve your garden's appearance while strengthening weed control.

Advantages of Blending Fabric and Mulch:

1. **Improved Weed Control:** The fabric functions as a shield, stopping weed seeds from sprouting and spreading through the ground. The addition of mulch on top of the fabric blocks sunlight and adds another physical barrier to further prevent weed growth.

2. **Enhanced Capture of Soil Moisture:** Mulch lowers evaporation, which aids in the retention of soil moisture. This impact is enhanced when landscape

39

fabric is added, guaranteeing that plants receive a steady flow of water.

3. **Control of Temperature:** Mulch helps to keep the soil's temperature steady by insulating it. Because it shields plant roots from intense heat or cold, this is very helpful in harsh weather situations.

4. **Visual Allure:** Mulch gives garden beds and landscapes a polished, expert appearance. It comes in a variety of forms, so you may pick one that goes well with your landscape design, including wood chips, bark, or beautiful stones.

Application Advice:

- **Select the Correct Mulch:** Choose a mulch type that complements the design and functional requirements of your landscaping. While inorganic mulches, like stones, provide long-lasting covers, organic mulches, like wood chips and bark, break down over time and enrich the soil with nutrients.

- **Stratification:** Mulch should be layered two to three inches thick over the landscaping fabric. To maintain consistent weed control and moisture retention, make sure the distribution is even.

- **Maintenance:** To keep the mulch layer looking and functioning properly, periodically inspect it and add more as needed.

When heavy-duty landscape fabric is combined with mulch, the result is a low-maintenance, aesthetically pleasing landscape that efficiently inhibits weed growth and retains soil moisture.

6.2 Edging: Establishing Clearly Defined Boundaries and Avoiding Fabric Overflow

Because it creates clear borders and keeps the cloth from rolling over or sliding, edging is a crucial part of landscape design.

Advantages of Edging:

- **Defined Borders:** Edging draws distinct, clean lines separating various landscape features, like lawns, garden beds, and walkways. This improves the landscape's general organization and aesthetic appeal.

- **Avoids Rollover of Fabric:** Edging stops the landscape cloth from moving or toppling over by tying off the edges. This makes sure the cloth stays where it is supposed to, which preserves its ability to suppress weeds and reduce erosion.

- **Decreased Upkeep:** By holding mulch and soil in place, edging helps lessen the frequency of upkeep and cleanup.

Types of Edging:

1. **Plastic Edging:** This flexible, lightweight edging is perfect for curving borders and is very simple to install.

2. **Metal Edging:** Sleek and contemporary, metal edging is long-lasting and durable. It works well with both curved and straight borders.

3. **Brick or Stone Edging:** These materials are incredibly durable and have a natural, timeless appeal. They are perfect for making borders that are aesthetically pleasing and more durable.

Installation Advice:

- **Correct Positioning:** Attach edging to the landscape fabric's edge, making sure it is securely fastened to the ground.
- **Securing the Fabric:** To keep the fabric from moving, overlap the edges with the edging material and fasten it with either staples or pins.
- **Consistent Height:** To preserve uniform borders and avoid trip hazards, make sure the edging is put at a consistent height.

When heavy-duty landscape fabric is used with edging, the landscape becomes more stable and attractive, resulting in garden areas that are well-defined and require little upkeep.

6.3 Pre-Emergent Herbicide: An Intensive Method for Difficult Weeds

When using landscape fabric, pre-emergent herbicides can be a useful tool for managing weeds that are difficult to eradicate.

Pre-emergent herbicides have the following advantages:

1. **Targeted Weed Control:** They stop weed seeds from sprouting and spreading. They work especially well against grasses and annual weeds that are difficult to control with cloth alone.

2. **Improved Weed Control:** Pre-emergent herbicides offer an extra layer of weed control when combined with landscape fabric, guaranteeing a weed-free landscape.

3. **Decreased Upkeep:** Pre-emergent herbicides save time and labor by reducing the need for manual weeding and herbicide applications by stopping weed germination.

Application Hints:

- **Time:** Before weed seeds sprout, usually in early spring or late fall, use pre-emergent herbicide. For optimal outcomes, adhere to the manufacturer's instructions.

- **Uniform Coverage:** Make sure the herbicide is

applied evenly throughout the entire area that has to be treated. To ensure even covering, use a sprayer or spreader.

- **Watering:** To assist the herbicide penetrate the soil and activate it, lightly water the area after application.

Pre-emergent herbicides and heavy-duty landscape fabric work together to give complete weed control that requires little upkeep and leaves landscapes free of weeds.

6.4 Overlap of Landscape Fabric: Guaranteeing Total Weed Control

Maintaining the integrity of the installation and guaranteeing total weed suppression depend on properly overlapping the landscape fabric.

Significance of Overlapping:

1. **Avoiding Weed Growth:** By overlapping the fabric's borders, weeds are prevented from growing via spaces left open to them. This is necessary to

keep up a constant barrier that prevents weed development.

2. **Longevity and Stability:** When the overlap is done correctly, the cloth is kept in place and doesn't move or separate. This guarantees stability and efficacy over the long run.

Overlapping Methods:

1. **6-Inch Overlap:** In most cases, a 6-inch overlap is enough to maintain stability and stop weed development. Using landscape staples, secure the fabric such that the edges overlap by a minimum of 6 inches.

2. **Automatic Overlap:** Apply a graduated overlap approach in places that have uneven or sloping terrain. Make sure that every piece of fabric overlaps the preceding one by a minimum of 6 inches by laying the fabric in an uneven manner. In addition to adding stability, this stops soil erosion.

3. **Double Layer:** Take into account applying a double layer of fabric in locations where weeds are a recurring issue. First, layer one with a 6-inch

overlap, and then layer two on top of the first, again with a 6-inch overlap. This offers further defense against weeds.

Installation Advice:

- **Safety First:** To keep the overlaps in place and stop them from shifting, use landscape staples or pins. Staples should be placed along the seams and edges every 12 to 24 inches.
- **Look for Any Gaps:** Examine the overlaps once the cloth has been laid to make sure there are no exposed soil areas or gaps. If necessary, adjust the fabric to ensure full coverage.

A continuous barrier against weeds is ensured by properly overlapping landscape cloth, which improves the installation's efficacy and endurance.

A thorough approach to landscape management is achieved by combining heavy duty landscape fabric with other products including mulch, edging, pre-emergent herbicides, and appropriate overlapping techniques. By improving weed suppression, moisture retention, erosion management, and aesthetic appeal, these techniques increase the fabric's performance and produce a landscape that is both aesthetically pleasing and well-maintained.

CHAPTER 7

UPKEEP ADVICE FOR DURABLE PERFORMANCE

7.1 Examining for Tears and Punctures: Immediate Damage Management

Heavy duty landscape fabric's long-term efficacy in weed control and soil protection depends on its integrity being preserved. It is ensured that the fabric will continue to function at its best by routinely checking it for tears and punctures.

Importance of Frequent Inspections:

1. **Preventing Weed Growth:** Weeds can enter a material through tears and punctures. Resolving these problems as soon as possible contributes to a landscape free of weeds.
2. **Preserving Soil Architecture:** Erosion of the soil and instability may result from damage to the fabric.

Frequent examinations aid in preventing erosion and preserving the soil's structure.

3. **Prolonging Fabric Lifespan:** You may prolong the life of the fabric and prevent the need for pricey replacements by spotting and fixing damage early.

Tips for Inspection:

Frequency: Check the fabric at least twice a year, ideally in the late fall and early spring. It could be essential to conduct further inspections following extreme weather or intensive gardening tasks.

Extensive Analysis: Examine the entire area that the fabric covers, being especially careful to inspect the edges, seams, and high-traffic areas. Keep an eye out for rips, tears, and punctures.

Repair Tools: Maintain a supply of repair supplies on hand, such as utility knives, fabric patches, and landscaping staples. Cover and secure any damaged areas with extra pieces of cloth or UV-resistant mending tape.

To guarantee that the fabric continues to effectively control weeds and protect the soil, it is imperative to conduct thorough and routine inspections in order to swiftly identify and treat any damage.

7.2 Restocking Mulch: Sustaining a Layer That Prevents Weeds

Mulch is essential for improving the performance of landscape fabric. To guarantee that mulch keeps its ability to inhibit weed growth and hold onto soil moisture, it must be kept at an appropriate layer.

Advantages of Maintaining Mulch:

1. **Improved Weed Control:** Sunlight is blocked by a thick layer of mulch, which stops weed seeds from sprouting and spreading. Adding mulch helps to keep this barrier of weed suppression in place.

2. **Enhanced Soil Humidity:** Mulch lowers evaporation, which aids in the retention of soil moisture. Restocking mulch on a regular basis guarantees steady moisture levels for strong plant development.

3. **Control of Temperature:** Because mulch is an

insulator, it shields plant roots from very cold temperatures. Keeping the mulch layer thick enough aids in regulating soil temperature all year round.

4. **Visual Allure:** Fresh mulch gives garden beds and landscapes a polished, well-kept appearance that improves their aesthetic appeal.

Advice for Restocking Mulch:

- **Regime:** Restock mulch ideally in the early spring, at least once a year. There might be a need for additional treatments following strong winds or rain.
- **Depth:** Keep a 2-3 inch mulch layer in place. Mulch that is applied in excess can suffocate plant roots, while too little diminishes its efficiency.
- **Equal Dispersion:** Be sure to cover the entire fabric with mulch by spreading it evenly over it. Mulch should not be piled up against plant stems as this might lead to decay and disease.

You can keep your mulch at a level that effectively suppresses weeds and improves the general health and aesthetics of your landscaping by topping it up on a regular basis.

7.3 Eliminating Superfluous Weeds: Handling Breakouts Above the Fabric

Even with the application of dense mulch and landscape fabric, weed outbreaks can still happen occasionally. Maintaining the general health and aesthetic appeal of the landscape requires swiftly addressing them.

Importance of Timely Weed Removal:

1. **Preventing Spread:** Pulling weeds out before they seed keeps them from growing a larger root system and from spreading.

2. **Maintaining Aesthetic Appeal:** A landscape free of weeds appears expertly cared for. Timely removal of weeds contributes to maintaining the aesthetic value of garden beds and pathways.

3. **Preserving the Health of Plants:** When it comes to sunlight, water, and nutrients, weeds compete with desired vegetation. Eliminating weeds guarantees that your plants get the nourishment they require to

flourish.

Tips for Weed Removal:

- **Hand Pulling:** Hand pulling is a useful technique for modest weed outbreaks. To stop regeneration, make sure you remove the entire root.
- **Weed Instruments:** For bigger or more deeply rooted weeds, use the proper weeding equipment, like a hoe or weeder. Take care not to rip the cloth.
- **Natural Weed Management:** For stubborn weeds, think about applying organic weed control techniques like boiling water or vinegar. These techniques are safe for plants and the environment.

You can keep up a beautiful, well-maintained environment that encourages the growth of desired plants by quickly controlling weed outbreaks.

7.4 Using UV-Treated Fabric and Appropriate Installation to Increase Fabric Lifespan

Choosing premium materials and making sure your heavy

duty landscaping fabric is installed correctly are key to extending its longevity.

Significance of UV-Treated Fabric:

1. **Sunlight Resistance:** UV-treated fabric is made to resist prolonged exposure to sunlight, which keeps it from deteriorating and lengthens its life.
2. **Durability:** UV-treated cloth resists weeds and protects soil for an extended period of time while retaining its structural integrity.
3. **Economic Viability:** UV-treated fabric may initially cost more, but over time, its longer lifespan makes it a more economical option.

Installing Properly:

- **Placement:** Make sure the fabric is laid out flat and overlapped correctly at the seams to avoid gaps and movement. Landscape staples or pins should be used at regular intervals to secure the fabric.
- **Preventing Weeping:** When installing the fabric, handle it gently to prevent rips and punctures. To cut

the fabric cleanly for planting or modifying, use sharp tools.

- **Securing and Edging:** To hold the fabric's edges and stop it from shifting or rolling over, use edging. This aids in preserving the position and efficacy of the fabric throughout time.

Maintenance Considerations:

- **Regular Inspections:** Check for wear or damage on a regular basis by conducting inspections. To keep the cloth functional, quickly mend any rips or punctures.
- **Correct Mulching:** To further prolong the fabric's life and shield it from physical harm and direct sunlight, make sure there is a sufficient layer of mulch covering it.

You can greatly increase the lifespan of your landscaping fabric and guarantee long-lasting performance and efficient weed suppression by using UV-treated fabric and according to recommended installation and maintenance procedures.

Heavy duty landscape fabric requires routine maintenance that includes looking for damage, keeping an adequate layer of mulch on the ground, pulling weeds quickly, and utilizing high-quality materials installed correctly. These procedures guarantee that the fabric keeps up its good work, offering enduring advantages for erosion control, soil moisture retention, and weed suppression.

CHAPTER 8

SELECTING THE IDEAL 6 FEET WIDE LANDSCAPE CLOTH

Making the proper landscape fabric choice is essential to the longevity and success of your landscaping work. Professionals and do-it-yourselfers both favor a 6 foot wide landscape cloth because it has several benefits. This chapter will discuss the advantages of utilizing 6 feet wide landscape fabric, how to determine how much fabric you'll need, the differences between pre-cut and custom-cut options, and where to get premium 6 feet wide heavy duty landscape fabric.

8.1 Benefits of a 6-foot-wide cloth: Reduced Seams and Simplified Installation

There are several advantages to selecting a 6 foot wide landscape fabric, all of which help to make the landscaping process go more smoothly and successfully.

Less Seams:

1. **Decreased Weed Penetration:** Less seams means less places where weeds could potentially get in. This improves the fabric's overall ability to inhibit weeds.
2. **Increased Sturdiness:** Seamless installations are typically more resilient and less likely to rip or move, extending the life of the fabric.
3. **Better Aesthetics:** Less seams result in a smoother, more uniform appearance; this is particularly crucial for places that are visible, such garden beds and pathways.

Simpler Installation:

- **Time Efficiency:** Broader cloth covers a larger area with each roll, requiring fewer pieces and less installation time.
- **Simplicity:** Installing a landscape is made easier for both professionals and do-it-yourselfers when fewer pieces need to be handled.
- **Consistency:** Greater coverage is ensured by larger

portions of fabric, which lowers the possibility of overlaps or gaps.

The fabric's 6 feet of width makes installation easier and offers excellent coverage and durability, which makes it a great option for a variety of landscaping tasks.

8.2 Estimating Fabric Requirements for Your Project: Width and Depth

You may prevent shortages and excess by purchasing the appropriate quantity of landscape fabric for your project by accurately assessing the amount required.

Assessing the Surface:

1. Establish the project's dimensions by taking measurements of the area's length and width. Measure each section of an irregularly shaped area after dividing it into smaller ones.
2. Take Overlaps Into Account: To ensure stability and successful weed suppression, add extra material for overlaps (6–12 inches each seam) when determining

the overall amount of fabric required.

Finding the Length of the Fabric:

- **Area Calculation:** To calculate the total square footage, multiply the length by the breadth of the area.

- **Fabric Width Consideration:** To calculate the required length, divide the total square footage by the fabric's width (6 feet). For instance, 50 linear feet of cloth measuring six feet wide will be required to cover 300 square feet.

- **Add Extra for wastage:** To accommodate for wastage and cutting errors, consider adding an additional 10% to your overall fabric required.

Precise computations assist you in placing the appropriate cloth order, guaranteeing effective coverage and reducing waste.

8.3 Examining Custom Cut vs. Pre-Cut Options for 6-Feet of Fabric

You can buy rolls of landscape fabric that are already cut, or you can purchase custom cut lengths. Every choice offers advantages and things to think about.

Pre-Cut Choices:

- **Convenience:** Pre-cut rolls are easily found in conventional lengths (such as 50 feet or 100 feet), which makes them practical for use right away.
- **Economically Sound:** Pre-cut rolls in bulk are frequently less expensive per foot than custom cuts.
- **Available:** Garden centers, home improvement stores, and internet merchants all stock pre-cut rolls.

Custom Cut Options:

- **Customization:** You can order custom cut fabric in the precise length you need for your project, which saves waste and guarantees a great fit.

- **Special Projects:** Custom cuts provide more flexibility and precision for projects with unusual dimensions or particular needs.
- **Decreased Seams:** The amount of seams can be reduced with custom cuts, particularly for large or asymmetrical regions.

The exact requirements, tastes, and financial constraints of your project will determine whether to use pre-cut or custom-cut fabric. Each of the solutions has unique benefits that can help your landscaping project succeed.

8.4 Where to Get Premium 6 feet Wide, Thick Landscape Fabric

By choosing a reputable supplier, you can be sure that the materials you buy for your project will be of the highest caliber.

Local Home Improvement businesses and Garden Centers:

- **Immediate Availability:** Local businesses

frequently have a range of landscape fabrics, so you may feel and examine the item before you buy it.

- **Expert Advice:** Employees in the store are able to offer suggestions and guidance on choosing the best cloth for your project.
- **Support Local Businesses:** Buying locally owned goods stimulates the community's economy and offers chances to establish rapport with suppliers.

Online retailers:

- **Extensive Variety:** Online marketplaces provide a huge assortment of landscape textiles, including various brands, weights, and compositions.
- **Customer Reviews:** Online evaluations and ratings give you information about the fabric's performance and quality, assisting you in making a wise choice.
- **Friendliness:** You may order from the convenience of your home, compare prices, and read product descriptions when you purchase online.

Specialty Suppliers:

- **High-Quality Materials**: Professional-grade landscaping fabrics that adhere to industry requirements are frequently offered by specialty suppliers.

- **Customization Options:** These vendors can provide UV-treated textiles, specially cut lengths, and other features that are specific to your project.

- **Expert Consultation:** Specialty suppliers can offer comprehensive product details and knowledgeable guidance, ensuring that you choose the ideal fabric for your requirements.

When buying landscape fabric, you can make sure you get high-quality supplies that improve the efficiency and longevity of your landscaping work by selecting reliable vendors.

Choosing the ideal 6 foot wide landscape fabric requires knowing its benefits, precisely estimating how much fabric you'll need, weighing your alternatives for pre-cut versus custom cut, and picking reliable retailers. You can guarantee that your landscaping projects will be successful and last a long time by making wise decisions at every stage.

CHAPTER 9

EXCEEDING WEED CONTROL: SURPRISING ADVANTAGES

Although weed control is the main application for heavy duty landscape fabric, its many other uses can improve many elements of your outdoor settings. This chapter looks at how landscape fabric may help control temperature, make play spaces better, improve kennels and pet runs, and even act as a firebreak.

9.1 Controlling Temperature: Sustaining Lower Soil Temperatures

Controlling Soil Temperature:

- **reflecting Properties:** The reflecting surfaces of high-quality landscaping fabrics can help deflect sunlight, which lowers soil temperatures.
- **Insulation:** The fabric can provide a more stable growing environment for plants by serving as a

barrier to shield the soil from sharp temperature swings.

Advantages for Plant Health:

- **Reduced Heat Stress:** Plants that are exposed to less heat stress, particularly in the summer, grow more healthily and produce more.
- **Moisture Retention:** Soils with a lower temperature have a tendency to hold onto moisture longer, which minimizes the need for frequent watering and preserves water supplies.

Plant health, water saving, and the sustainability of gardens as a whole can all be greatly enhanced by maintaining cooler soil temperatures through the use of landscape fabric.

9.2 Play Areas: Establishing a Clean and Safe Environment for Kids

Safety Enhancements:

- **Soft Underlayment:** Landscape fabric adds an extra layer of cushioning beneath playground mulch or rubber chips, increasing play areas' safety by lowering the possibility of falls-related injuries.
- **Organicness:** By acting as a barrier and preventing dirt and muck from combining with the surface material, the fabric helps to maintain play areas hygienic and clean.

Benefits for Maintenance:

- **Weed Suppression:** The fabric successfully suppresses weeds, requiring less maintenance and guaranteeing a weed-free, hygienic play area.
- **Easy Cleaning:** By keeping dirt from sinking into the soil, landscape fabric makes it easier to clean play areas and preserve a neat appearance.

Play spaces with landscape fabric offer kids a cleaner and safer atmosphere while requiring less upkeep from parents and other caregivers.

9.3 Pet Runs and Kennels: Reducing Mud and Upholding Sanitation

Hygiene and Cleanliness:

- **Mud Reduction:** Landscape fabric reduces mud formation by forming a barrier between the soil and the surface material, making kennels and pet runs cleaner.
- **Simple Upkeep:** Because the cloth keeps waste and filth from seeping into the ground, it makes cleaning pet areas easier and promotes better sanitation.

Benefits for Pet Health:

- **Reduced Pests:** Less dirt and debris in the environment means fewer pests and parasites that could injure pets.
- **Solace:** Pet comfort is increased on a dry, clean surface, which improves general health and wellbeing.

Pet owners will find landscape fabric to be an excellent

choice as it can significantly minimize insect problems, increase pet comfort, and improve hygiene in pet runs and kennels.

9.4 Building a Barrier to Slow the Spread of Flames: Firebreaks

Preventing Fires:

- **Building Barriers:** Firebreaks are obstacles intended to impede or completely put out the spread of flames. They can be made with landscape fabric. This is especially helpful in places where wildfires are common.
- **Vegetable Management:** Landscape fabric lowers the amount of flammable material accessible and hence the likelihood of a fire spreading by inhibiting the growth of combustible vegetation.

Strategic Implementation:

Placement: To enhance their effectiveness, firebreaks should be strategically positioned around buildings, along property lines, and in densely vegetated regions.

Maintenance: To achieve the best possible fire prevention, firebreaks must be regularly maintained. This includes keeping the fabric free of debris and guaranteeing its integrity.

By adding a second line of defense against wildfires, landscape fabric firebreaks preserve property and improve safety in areas where wildfires are common.

Heavy strength landscaping fabric has many unanticipated uses beyond just controlling weeds; these uses can greatly improve outdoor areas. Landscape fabric shows to be a useful and adaptable solution for a variety of applications by controlling soil temperatures, establishing safe play spaces, enhancing pet runs and kennels, and acting as firebreaks. Accepting these extra advantages can result in safer surroundings, healthier plants, and better landscape management all around.

CHAPTER 10

USING HEAVY DUTY LANDSCAPE FABRIC TO GO GREEN

Selecting ecologically responsible solutions for heavy duty landscape fabric can make a big difference in a time when many landscaping projects prioritize sustainability. This chapter examines the range of environmentally friendly options for landscape fabric, such as products manufactured from recycled materials, fabrics that allow water to drain through them, fabrics that decompose naturally over time, and sustainable landscaping techniques that encourage the wise use of landscape fabric.

10.1 Eco-Friendly Choices for Recycled Content in Landscape Fabric

Eco-Friendly Manufacturing:

Recycled Materials: Plastic bottles and other post-consumer waste are recycled to make a large number of landscape fabrics these days. This helps keep garbage out of landfills and lowers the need for fresh raw materials.

Production Procedures: These environmentally friendly textiles are made using sustainable manufacturing techniques that reduce their negative effects on the environment. This covers the use of non-toxic dyes and treatments as well as energy-efficient production techniques.

Environmental Benefits:

- **Reduced Carbon Footprint:** Making landscape fabric from recycled materials considerably reduces the product's carbon footprint. This is because recycling uses less energy and produces fewer emissions than creating new materials.
- **Resource Conservation:** By selecting textiles with recycled content, you support a more sustainable use of materials by assisting in the conservation of natural resources.

Choosing landscape fabric with recycled content is a great method to protect the soil and effectively manage weeds in your landscaping projects while also being environmentally conscious.

10.2 Options for Permeable Fabrics: Letting Water Pass Through

Managing Water:

- **Drainage:** The goal of permeable landscape fabrics is to keep weeds out of the garden while still allowing water to flow through. This makes sure that irrigation and rainfall can get to the soil, which encourages the growth of healthy plants.
- **Control of erosion**: Permeable materials aid in preventing soil erosion and surface runoff by letting water pass through, preserving the integrity of your landscape.

Health of the Soil:

- **Aeration:** Additionally, by facilitating air exchange between the soil and the atmosphere, these materials help to improve soil aeration, which is essential for the health of the roots and microbial activity.

- **Wetness Equilibrium:** In order to prevent waterlogging and guarantee that plants receive the proper amount of water, permeable materials aid in maintaining the ideal moisture balance in the soil.

Permeable landscape fabric is a sustainable option for many landscaping applications because it promotes healthy soil and efficient water management.

10.3   Biodegradable   Fabric: Taking   Future Disintegration Into Account

Short-Term Projects:

- **Temporary Solutions:** For short-term projects or circumstances where the fabric is only required for a brief time, biodegradable landscape fabrics are perfect. These textiles are meant to decompose organically over time, so removal and disposal are not necessary.
- **Green:** Biodegradable textiles, which are made from natural materials like jute, coir, or other plant fibers,

break down into organic matter and replenish the soil instead of piling up in landfills.

Environmental Impact:

- **Reduced Waste:** Compared to typical landscape fabrics, biodegradable fabrics produce a lot less plastic waste.
- **Soil Enrichment:** As biodegradable textiles break down, they enrich the soil with organic content, enhancing its fertility and structure.

In keeping with sustainable landscaping techniques, taking into account biodegradable landscape fabric is a progressive move that effectively suppresses weeds while promoting soil health and cutting down on waste.

10.4 Conscientious Fabric Use in Sustainable Landscaping: A Responsible Approach

Integrated Landscaping:

- **Companion Planting:** To naturally reduce weeds

and encourage biodiversity, combine the use of landscape fabric with companion planting techniques.

- **Native Plants:** Include native plants that are acclimated to the area well to minimize the need for chemical inputs and excessive watering.

Responsibly Use:

- **Correct Installation:** Make sure landscape fabric is installed correctly to enhance its longevity and efficacy. This entails utilizing the proper anchoring techniques, overlapping seams, and fastening edges.
- **Minimal Intervention:** Rather of covering the entire landscape, use landscape fabric sparingly where it will be most helpful, such as around plants that require a lot of care or in places that are very susceptible to weed growth.

Upkeep and Disposal:

- **Regular Inspection:** Check the fabric on a regular basis for

-

- harm and take quick action to fix any problems to prolong its life.

- **Green Waste Removal:** When a fabric's useful life is coming to an end, choose biodegradable materials that will naturally break down without affecting the environment, or think about recycling possibilities if any are available.

A healthier, more resilient landscape is promoted by using sustainable landscaping techniques, which guarantee that the usage of landscape fabric is in line with more general environmental objectives.

The environmental advantages of your landscaping projects can be further increased by utilizing heavy duty landscape fabric in conjunction with sustainable and eco-friendly solutions. You can accomplish successful weed control and soil protection while advancing more general sustainability goals by selecting fabrics with recycled content, permeable materials, and biodegradable materials, as well as by putting responsible usage and maintenance techniques into practice. Adopting these eco-friendly solutions and methods makes landscaping more ecologically responsible and sustainable.

ABOUT THE AUTHOR

Prolific writer Terra Firma Royce is renowned for his perceptive examination of the natural world and how it connects to human experience. Royce, whose name suggests earthiness and firmness, writes extensively on topics like sustainability, ecology, and the close relationships that exist between humans and their surroundings.

Style and Literary Themes

The impact of human activities on the environment and environmental stewardship are major themes throughout Royce's literary works. His writing style is distinguished by a combination of moving observations on the brittleness and resiliency of nature with poetic descriptions of landscapes and ecosystems. By weaving tales together, Royce encourages readers to consider their part in protecting the natural environment by highlighting its fragility as well as its beauty.

Influence of Philosophy

In addition to his literary accomplishments, Terra Firma Royce is a respected authority on environmental philosophy. He supports a comprehensive strategy for conservation that recognizes the interdependence of all living things. His articles frequently address moral conundrums pertaining to environmental deterioration and people's ethical obligations to the next generation.

History and Significance

Because of its capacity to cause readers to reevaluate their relationship with the environment, Royce's work has won praise from critics. His support of ecological consciousness and sustainable behaviors is still relevant to a wide range of people, from legislators looking for fresh approaches to the world's environmental problems to ardent nature lovers.

www.ingramcontent.com/pod-product-compliance
Lightning Source LLC
Chambersburg PA
CBHW071946210526
45479CB00002B/835